## INDIA STATE HUNGER INDEX

# Comparisons of Hunger Across States

Purnima Menon, Anil Deolalikar, Anjor Bhaskar

An advance copy of this report was released at a workshop on October 14, 2008, that also launched the Global Hunger Index 2008. This revised version incorporates feedback received at the workshop as well as written feedback received from a number of other readers.

**Washington, D.C., Bonn, and Riverside • February 2009**

IFPRI INTERNATIONAL FOOD POLICY RESEARCH INSTITUTE
*sustainable solutions for ending hunger and poverty*
Supported by the CGIAR

welt hunger hilfe

UNIVERSITY OF CALIFORNIA UC RIVERSIDE

# Acknowledgments

This work was supported by Welthungerhilfe through a grant to the International Food Policy Research Institute. We are grateful to the following individuals for valuable comments and feedback: Doris Wiesmann, Marie Ruel, Klaus von Grebmer, and other colleagues at IFPRI and Welthungerhilfe. We also thank the many readers of the advance version of this report who sent us their valuable feedback.

# Table of Contents

Legend
30 or more Extremely Alarming
20.0-29.9 Alarming
10.0-19.9 Serious
ISHI not estimated
Jammu & Kashmir
Himachal Pradesh
Chandigarh
Punjab
Uttarakhand
Haryana
Delhi
Rajasthan
Uttar Pradesh
Sikkim
Arunachal Pradesh
Assam
Nagaland
Meghalaya
Manipur
Bihar
Jharkhand
Tripura
Mizoram
West Bengal
Gujarat
Madhya Pradesh
Chhattisgarh
Daman & Diu
Dadra & Nagar Haveli
Orissa
Maharashtra
Andhra Pradesh
Karnataka
Goa
Andaman & Nicobar
Lakshadweep
Puducherry
Tamil Nadu
Kerala

# Key Findings

- India's Global Hunger Index (GHI) 2008 score is 23.7, which gives it a rank of 66th out of 88 countries. This score indicates continued poor performance at reducing hunger in India.
- The India State Hunger Index (ISHI) 2008 was constructed in a similar fashion as the GHI 2008 to allow for comparisons of states within India and for comparisons of Indian states to GHI 2008 scores and ranks for other countries.
- The ISHI 2008 score was estimated for 17 major states in India, covering more than 95 percent of the population of India.
- ISHI 2008 scores for Indian states range from 13.6 for Punjab to 30.9 for Madhya Pradesh, indicating substantial variability among states in India. Punjab is ranked 34th when compared with the GHI 2008 country rankings, and Madhya Pradesh is ranked 82nd.
- All 17 states have ISHI scores that are significantly worse than the "low" and "moderate" hunger categories. Twelve of the 17 states fall into the "alarming" category, and one—Madhya Pradesh—falls into the "extremely alarming" category.
- ISHI scores are closely aligned with poverty, but there is little association with state-level economic growth. High levels of hunger are seen even in states that are performing well from an economic perspective.
- Inclusive economic growth and targeted strategies to ensure food sufficiency, reduce child mortality, and improve child nutrition are urgent priorities for all states in India.

# Introduction

Food and good nutrition are basic human needs, and this is recognized in the first Millennium Development Goal (MDG)—the eradication of extreme poverty and hunger. Developing sound ways to monitor progress toward the eradication of hunger is vital to productive global and national policy discussions about hunger. The Global Hunger Index (GHI) is one approach to measuring and tracking progress on hunger and enabling widespread discussion about the reasons for and consequences of hunger. The GHI was developed by the International Food Policy Research Institute (IFPRI) in 2006 (Wiesmann et al. 2006) as a means of capturing three interlinked dimensions of hunger—inadequate consumption, child underweight, and child mortality.

Although hunger is most directly manifested in inadequate food intake, over time inadequate food intake and a poor diet, especially in combination with low birth weights and high rates of infection, can result in stunted and underweight children. The most extreme manifestation of continued hunger and malnutrition is mortality. The Global Hunger Index recognizes the interconnectedness of these dimensions of hunger and captures the performance of all three of them. The index has been an effective advocacy tool that has brought the issue of global and national hunger to the fore in policy debates, especially in developing countries. The ranking of nations on the basis of their index scores has been a powerful tool to help focus attention on hunger, especially for countries like India that underperform on hunger and malnutrition relative to their income levels.

India has consistently ranked poorly on the Global Hunger Index. The Global Hunger Index 2008 (von Grebmer et al. 2008) reveals India's continued lackluster performance at eradicating hunger; India ranks 66th out of the 88 developing countries and countries in transition for which the index has been calculated. It ranks slightly above Bangladesh and below all other South Asian nations.[1] It also ranks below several countries in Sub-Saharan Africa, such as Cameroon, Kenya, Nigeria, and Sudan, even though per capita income in these Sub-Saharan African countries is much lower than in India (Table 1).

[1] India's slightly better performance relative to Bangladesh is entirely due to better access to food in India relative to Bangladesh, which in turn is a consequence of India's higher agricultural productivity. On the other two components of the GHI—child underweight and child mortality—India ranks below Bangladesh. Indeed, India's child underweight rates are among the highest in the world.

**TABLE 1—GDP PER CAPITA IN RELATION TO SCORES ON THE GLOBAL HUNGER INDEX 2008**

| Country | GHI 2008 | GDP per capita[a] |
|---|---|---|
| Nigeria | 18.4 | 1,977 |
| Cameroon | 18.7 | 2,124 |
| Kenya | 19.9 | 1,535 |
| Sudan | 20.5 | 2,088 |
| **India** | **23.7** | **2,753** |

Source: World Bank 2007 and von Grebmer et al. 2008.
[a] Gross domestic product (GDP) dollar estimates at purchasing power parity (PPP) per capita.

# A State-Level Hunger Index for India

With more than 200 million food-insecure people (FAO 2008), India is home to the largest number of hungry people in the world. Although the central government has given attention to hunger and undernutrition, India's states play important roles in planning and executing development programs. Unpacking the hunger index at the level of the state is a useful tool for building awareness of the levels of hunger in the states and disparities among them. In addition, findings concerning the different relative contributions of the underlying components of the hunger index across different Indian states can help inform discussions about the drivers of hunger in different state contexts.

The availability of national household surveys in India that capture the three underlying dimensions of the Global Hunger Index at the state level makes it possible to develop a state hunger index for India. The state hunger index is calculated using a procedure similar to that used for calculating the GHI scores of countries. Hunger and malnutrition in Indian states can thus be compared with hunger and malnutrition in other countries.

The overall objective of the development of an India State Hunger Index is to focus attention on the problem of hunger and malnutrition at the state and central levels in India through the development of an index that enables comparisons within India, and globally.

# Data and Methods

The India State Hunger Index is based on the same underlying variables as the Global Hunger Index. These variables are

- the proportion of population that does not consume an adequate level of calories;
- the proportion of underweight children under five years of age; and
- the mortality rate among children under five years of age, expressed as the percentage of children born alive who die before they reach the age of five.

We use two data sources to estimate the India State Hunger Index: the third round of the National Family Health Survey (2005–06) for India (referred to as the NFHS-III data)[2] and the 61st round of the National Sample Survey (NSS) data from 2004–05.[3]

The India State Hunger Index is calculated and presented for 17 major states in the country. These states cover 95 percent of the country's population, according to the 2001 census in India. Although the NFHS-III has a sample size large enough to yield representative estimates of the child underweight and mortality rates even for the smaller states and union territories in the country, the sample size of the NSS 61st round is insufficient for estimating undernourishment rates in these places. Therefore, we restricted our sample to those states for which the NSS yields precise state-level estimates.

Like the GHI, the India State Hunger Index is computed by averaging the three underlying components of the hunger index. This approach to combining the variables offers simplicity and transparency. Previous efforts to develop other weighting approaches for combining these variables revealed that, for advocacy purposes, a simple and transparent approach was preferable to more complex approaches that weighted the variables differently. Additionally, other weighting approaches did not offer any empirical advantages over the simpler approach (Weismann 2006).

## UNDERNOURISHED POPULATION BASED ON CALORIE CONSUMPTION

The proportion of the population that is undernourished based on calorie consumption was estimated using unit-level food consumption data from the 61st round of the NSS, conducted in 2004–05. The NSS obtained household consumption data on more than 225 individual foods; these consumption data were converted to calories using food-to-calorie conversion factors reported by the National Sample Survey Organization (NSSO 2007). Allowances for calories from meals eaten outside the home were made using

[2] The National Family Health Survey III is the third in a series of such surveys. It was conducted by the International Institute for Population Sciences (IIPS), Mumbai, which acted as the nodal agency. The survey collected data on 51,555 children under 5 years of age. Technical support was provided by Macro International, and the U.S. Agency for International Development (USAID) provided financial assistance for the study.

[3] The National Sample Survey Organization (NSSO) conducts surveys on various socioeconomic issues annually. The 61st round of the NSS was conducted between July 2004 and June 2005. The Household Consumer Expenditure Survey, the seventh quinquennial survey on the subject, had a large sample consisting of 79,298 rural and 45,346 urban households in all the states and union territories of India.

the procedures suggested and followed by NSSO (2007). We include further details on these calculations in Appendix 1 of this report.

The GHI 2008 (von Grebmer et al. 2008) reports calorie-based undernutrition for India to be 20 percent, based on the undernourishment cutoff of 1,820 kilocalories (kcals) per person of the Food and Agriculture Organization of the United Nations (FAO).[4] The FAO estimate of calorie undernutrition is based on data on food availability, collected from national food balance sheets. As such, it is likely to differ from calorie undernutrition estimates that are based on consumption data obtained from household consumption expenditure surveys, such as the NSS (Smith and Wiesmann 2007). In the case of India, the use of 1,820 kcals per person per day as the cutoff yields a calorie undernutrition rate of 34 percent, which is substantially larger than the 20 percent reported by the FAO and used in the estimation of the Global Hunger Index 2008. Given that an important goal of this study was to ensure comparability of the India State Hunger Index with the Global Hunger Index, we use a calorie cutoff—1,632 kcals per person per day—that yields a national calorie undernutrition rate of exactly 20 percent.

[4] It should be noted that the calorie norms for poverty used within India differ significantly from the FAO-recommended cutoffs for undernourishment. Dandekar and Rath (1971) used a norm of 2,250 kcals per day per person to set a poverty line for India. A task force of the Indian Planning Commission subsequently revised this calorie norm to 2,400 in rural areas and 2,100 in urban areas (the difference being attributed to the lower rates of physical activity in the urban areas; GOI 1979). These are the calorie norms that underlie the official poverty line currently in use by the Government of India. To complicate matters further, the NSSO uses a daily calorie norm of 2,700 kcals per consumer unit (not per capita) (NSSO 2007).

It should be noted that calorie consumption alone is a conceptually inadequate measure of hunger. Without data on physical activity and calories expended, it is difficult to truly judge if an individual is undernourished or not. Nevertheless, it is a common indicator of hunger and is used widely in food security and hunger indexes, including the GHI. We recognize that there have been substantial debates on the use of this indicator in India (Dev 2005), but we retain it given our stated goal of ensuring comparability with the GHI.

## CHILD UNDERWEIGHT

The proportion of underweight among children under five years of age was estimated at the state level using unit-level data from the NFHS-III data set (available at www.measuredhs.com). We used the World Health Organization (WHO) 2006 international growth reference and NFHS-recommended sample survey weights to estimate the proportion of children in each state whose weight-for-age was less than two standard deviations below the WHO reference.

## CHILD MORTALITY

We use the under-five mortality rates at the state level as reported in the NFHS-III report (IIPS 2007, 187). The mortality rate is expressed as the percentage of live-born children who do not survive past age 59 months.

# Results

## CURRENT STATUS AND RANKING OF STATES WITHIN INDIA

Table 2 presents the India State Hunger Index, as well as its underlying components, for the 17 major states in the country. The severity of hunger is reported in Table 3, and the ranking of the 17 states by the India State Hunger Index is shown in Figure 1. The classification of the severity of hunger is from the Global Hunger Index 2008.

**TABLE 2—THE INDIA STATE HUNGER INDEX AND ITS UNDERLYING COMPONENTS**

| State | Prevalence of calorie under–nourishment (%) | Proportion of underweight among children <5 years (%) | Under-five mortality rate (deaths per hundred) | India State Hunger Index score | India State Hunger Index rank |
|---|---|---|---|---|---|
| Punjab | 11.1 | 24.6 | 5.2 | 13.63 | 1 |
| Kerala | 28.6 | 22.7 | 1.6 | 17.63 | 2 |
| Andhra Pradesh | 19.6 | 32.7 | 6.3 | 19.53 | 3 |
| Assam | 14.6 | 36.4 | 8.5 | 19.83 | 4 |
| Haryana | 15.1 | 39.7 | 5.2 | 20.00 | 5 |
| Tamil Nadu | 29.1 | 30.0 | 3.5 | 20.87 | 6 |
| Rajasthan | 14.0 | 40.4 | 8.5 | 20.97 | 7 |
| West Bengal | 18.5 | 38.5 | 5.9 | 20.97 | 8 |
| Uttar Pradesh | 14.5 | 42.3 | 9.6 | 22.13 | 9 |
| Maharashtra | 27.0 | 36.7 | 4.7 | 22.80 | 10 |
| Karnataka | 28.1 | 37.6 | 5.5 | 23.73 | 11 |
| Orissa | 21.4 | 40.9 | 9.1 | 23.80 | 12 |
| Gujarat | 23.3 | 44.7 | 6.1 | 24.70 | 13 |
| Chhattisgarh | 23.3 | 47.6 | 9.0 | 26.63 | 14 |
| Bihar | 17.3 | 56.1 | 8.5 | 27.30 | 15 |
| Jharkhand | 19.6 | 57.1 | 9.3 | 28.67 | 16 |
| Madhya Pradesh | 23.4 | 59.8 | 9.4 | 30.87 | 17 |
| **India** | **20.0** | **42.5** | **7.4** | **23.30** | |

Note: The India State Hunger Index represents the index calculated using a calorie undernourishment cutoff of 1,632 kcals per person per day to allow for comparison of the India State Hunger Index with the Global Hunger Index 2008. The ISHI score for India using this cutoff is 23.3 and corresponds more closely with the GHI 2008 score for India of 23.7 than any other calorie cutoff.

Sources: Calorie undernourishment: IIPS 2007; child underweight: IIPS 2007 and authors' calculations; under-five mortality rate: NSSO 2007 and authors' calculations.

**TABLE 3—SEVERITY OF INDIA STATE HUNGER INDEX, BY STATE**

| <4.9 (low) | <5.0–9.9 (moderate) | 10.0–19.9 (serious) | 20.0–29.9 (alarming) | 30.0 or more (extremely alarming) |
|---|---|---|---|---|
| None | None | Punjab | Haryana | Madhya Pradesh |
| | | Kerala | Tamil Nadu | |
| | | Andhra Pradesh | Rajasthan | |
| | | Assam | West Bengal | |
| | | | Uttar Pradesh | |
| | | | Maharashtra | |
| | | | Karnataka | |
| | | | Orissa | |
| | | | Gujarat | |
| | | | Chhattisgarh | |
| | | | Bihar | |
| | | | Jharkhand | |

Note: The categorization of states is done using the same cutoffs for severity as the Global Hunger Index 2008. India's GHI 2008 score of **23.7** places it in the "alarming" category.

Sources: Table 2 and von Grebmer et al. 2008.

**FIGURE 1—RANKING OF STATES WITHIN INDIA FROM LOWEST TO HIGHEST INDIA STATE HUNGER INDEX SCORES**

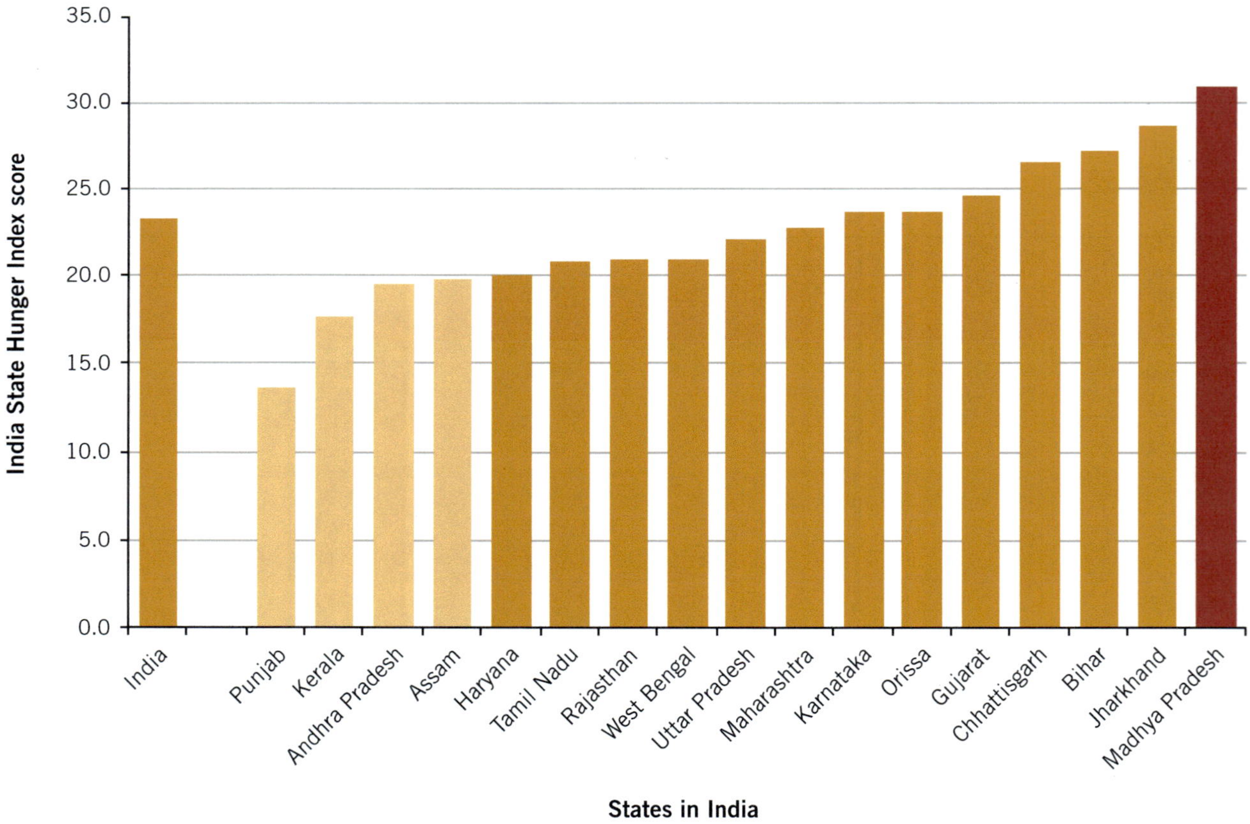


Source: See Table 2.

It is disconcerting that not a single state in India falls in the "low hunger" or "moderate hunger" category defined by the GHI 2008. Instead, most states fall in the "alarming" category, with one state—Madhya Pradesh—falling in the "extremely alarming" category. Four states—Punjab, Kerala, Andhra Pradesh, and Assam—fall in the "serious" category. The map of the India State Hunger Index in Figure 2 shows that the bulk of Indian states for which the hunger index was estimated are in the "alarming" category.

**FIGURE 2—MAP OF THE INDIA STATE HUNGER INDEX, BY SEVERITY**

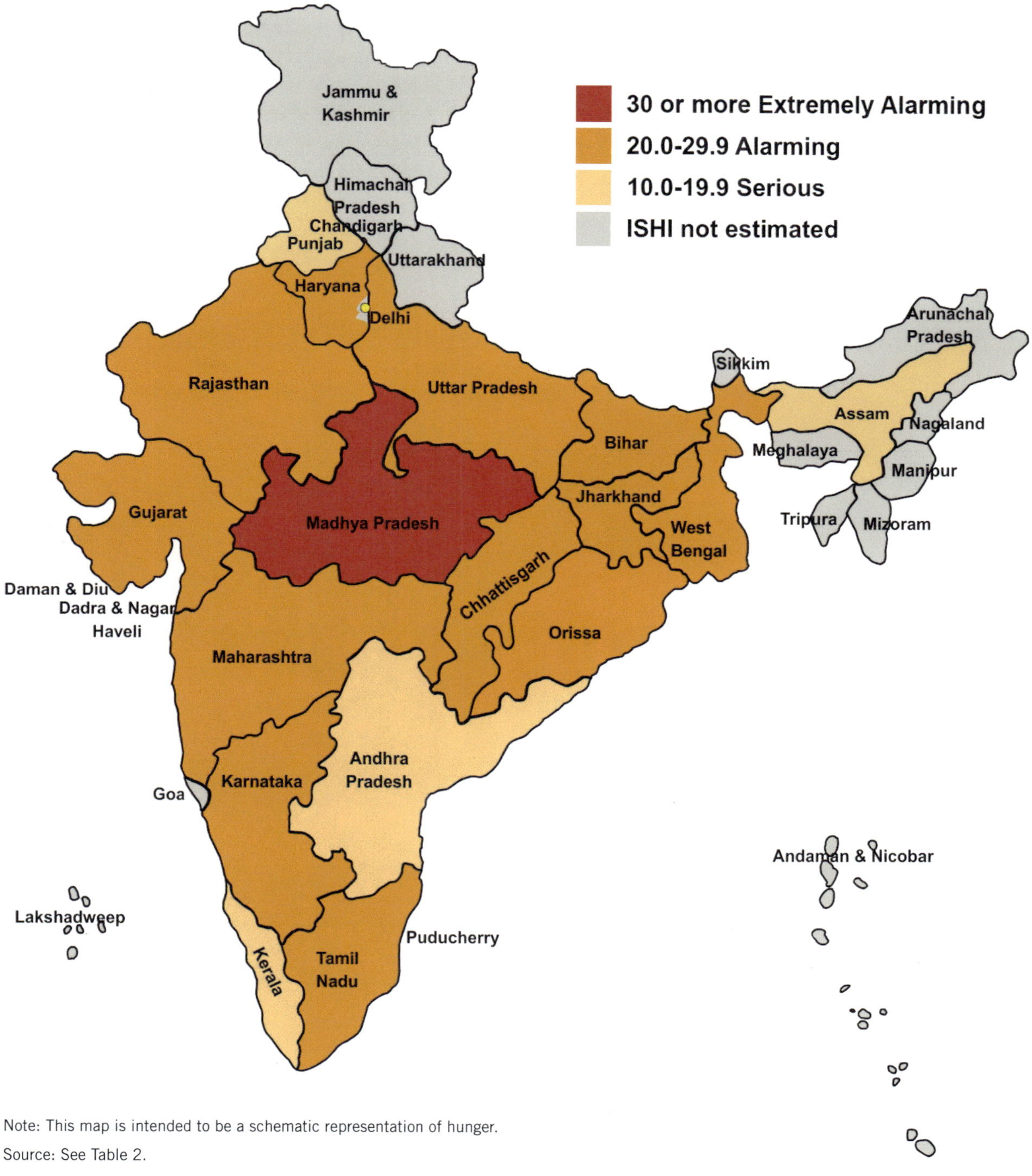


Note: This map is intended to be a schematic representation of hunger.

Source: See Table 2.

## COMPARISON OF INDIAN STATES WITH OTHER COUNTRIES

Table 4 shows the position of the 17 Indian states relative to the countries for which the Global Hunger Index 2008 is reported. India's rank on the GHI 2008 is 66th; the ranks of the different states in relation to the GHI range from 34th for the state of Punjab (whose ISHI score places it between Nicaragua and Ghana) to 82nd for Madhya Pradesh (whose ISHI score places it between Chad and Ethiopia). Ten of the 17 states have an ISHI rank that is above India's (66th)—these states are relative outperformers (at least relative to the Indian average). Even the best-performing state in India, however—Punjab—ranks below such countries as Gabon, Honduras, and Vietnam.

It is useful to examine the contributions of the underlying dimensions to the overall hunger index. For the majority of states, child underweight is responsible for the largest variation between states (Figure 3). In addition, for most states, overall scores are high because of particularly high child underweight rates. When compared with the majority of states, the contribution of low calorie consumption levels to the hunger index is higher for Kerala and Tamil Nadu, as well as, to a lesser extent, for Maharashtra and Karnataka. The contribution of child mortality to the hunger index scores, however, is relatively small and less variable across all the states when compared with the contributions of child underweight and calorie undernourishment.

**FIGURE 3—CONTRIBUTION OF UNDERLYING COMPONENTS OF THE INDIA STATE HUNGER INDEX TO OVERALL SCORES**

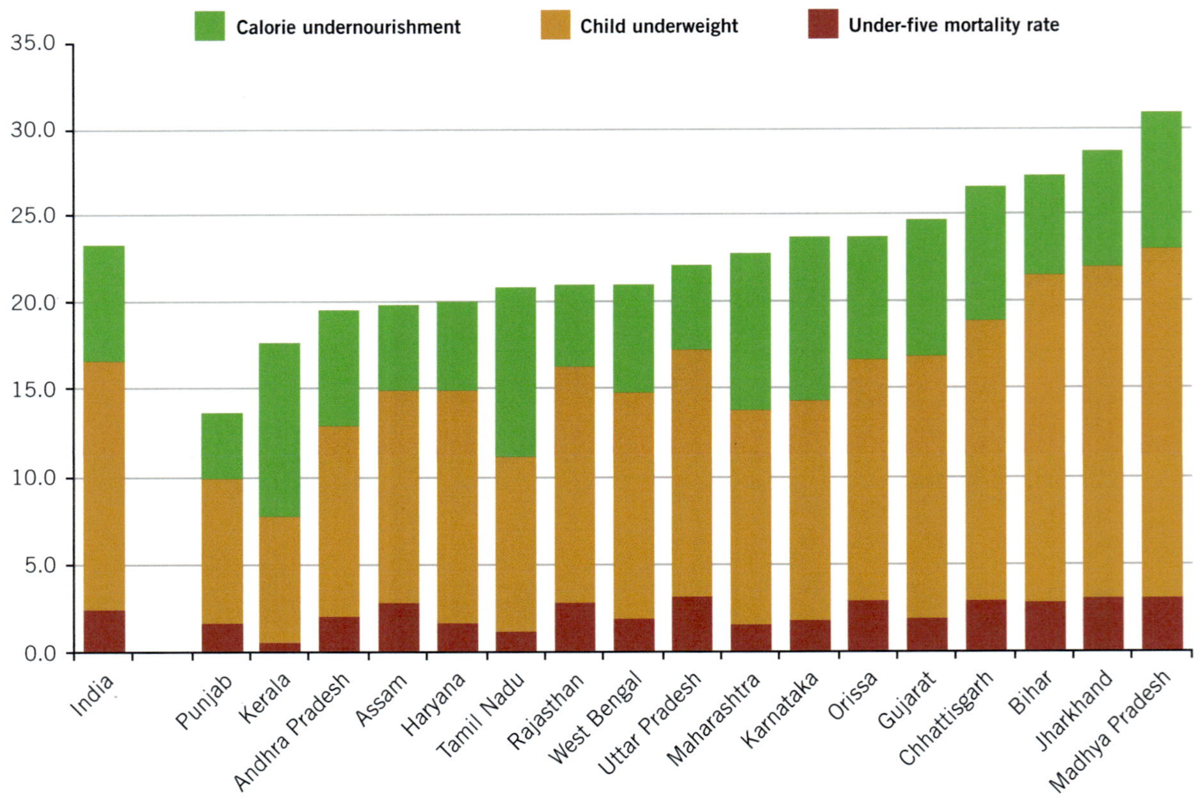


Source: See Table 2.

## TABLE 4—COMPARISON OF INDIAN STATES WITH COUNTRIES OUTSIDE INDIA, BASED ON THE GHI 2008

Note: Green shaded cells represent Indian States

| GHI Rank | Country/State | Hunger Index | GHI Rank | Country/State | Hunger Index | GHI Rank | Country/State | Hunger Index |
|---|---|---|---|---|---|---|---|---|
| | Argentina | <5 | 16 | Fiji | 7.3 | 58 | Lao PDR | 20.6 |
| | Belarus | <5 | 17 | Suriname | 7.5 | 59 | Djibouti | 20.9 |
| | Bosnia and Herzegovina | <5 | 18 | Gabon | 7.6 | 60 | Guinea | 20.9 |
| | Brazil | <5 | 19 | Venezuela | 7.7 | | Tamil Nadu | 20.9 |
| | Bulgaria | <5 | 20 | Paraguay | 7.9 | | Rajasthan | 21.0 |
| | Chile | <5 | 21 | Guyana | 8.6 | | West Bengal | 21.0 |
| | Costa Rica | <5 | 22 | Panama | 8.9 | 61 | Pakistan | 21.7 |
| | Croatia | <5 | 23 | Thailand | 9.9 | 62 | Malawi | 21.8 |
| | Ecuador | <5 | 24 | Armenia | 10.2 | | Uttar Pradesh | 22.1 |
| | Egypt, Arab Rep. | <5 | 25 | Azerbaijan | 10.4 | 63 | Rwanda | 22.3 |
| | Estonia | <5 | 26 | Uzbekistan | 11.2 | | Maharashtra | 22.8 |
| | Iran, Islamic Rep. | <5 | 27 | Indonesia | 11.3 | 64 | Cambodia | 23.2 |
| | Jordan | <5 | 28 | Honduras | 11.4 | 65 | Burkina Faso | 23.5 |
| | Kazakhstan | <5 | 29 | Bolivia | 11.7 | **66** | **India** | **23.7** |
| | Kuwait | <5 | 30 | Dominican Republic | 12.0 | | Karnataka | 23.7 |
| | Kyrgyz Republic | <5 | 31 | Mongolia | 12.1 | | Orissa | 23.8 |
| | Latvia | <5 | 32 | Vietnam | 12.6 | 67 | Zimbabwe | 23.8 |
| | Lebanon | <5 | 33 | Nicaragua | 12.8 | 68 | Tanzania | 24.2 |
| | Libya | <5 | | Punjab | 13.6 | 69 | Haiti | 24.3 |
| | Lithuania | <5 | 34 | Ghana | 13.9 | | Gujarat | 24.7 |
| | Macedonia | <5 | 35 | Philippines | 14.0 | 70 | Bangladesh | 25.2 |
| | Mexico | <5 | 36 | Lesotho | 14.3 | 71 | Tajikistan | 25.9 |
| | Romania | <5 | 37 | Namibia | 14.3 | 72 | Mozambique | 26.3 |
| | Russian Federation | <5 | 38 | Guatemala | 14.6 | | Chhattisgarh | 26.6 |
| | Saudi Arabia | <5 | 39 | Myanmar | 15.0 | 73 | Mali | 26.9 |
| | Serbia and Montenegro | <5 | 40 | Sri Lanka | 15.0 | | Bihar | 27.3 |
| | Slovak Republic | <5 | 41 | Benin | 15.1 | 74 | Guinea-Bissau | 27.5 |
| | Syrian Arab Republic | <5 | 42 | Côte d'Ivoire | 15.3 | 75 | Central African Republic | 28.0 |
| | Tunisia | <5 | 43 | Senegal | 15.4 | | Jharkhand | 28.7 |
| | Turkey | <5 | 44 | Uganda | 17.1 | 76 | Madagascar | 28.8 |
| | Ukraine | <5 | 45 | Gambia, The | 17.3 | 77 | Comoros | 29.1 |
| | Uruguay | <5 | 46 | Mauritania | 17.6 | 78 | Zambia | 29.2 |
| 1 | Mauritius | 5.0 | | Kerala | 17.6 | 79 | Angola | 29.5 |
| 2 | Jamaica | 5.1 | 47 | Swaziland | 17.7 | 80 | Yemen, Rep. | 29.8 |
| 3 | Moldova | 5.4 | 48 | Botswana | 17.9 | 81 | Chad | 29.9 |
| 4 | Cuba | 5.5 | 49 | Togo | 18.2 | | Madhya Pradesh | 30.9 |
| 5 | Peru | 5.6 | 50 | Timor-Leste | 18.4 | 82 | Ethiopia | 31.0 |
| 6 | Trinidad and Tobago | 5.9 | 51 | Nigeria | 18.4 | 83 | Liberia | 31.8 |
| 7 | Algeria | 6.0 | 52 | Cameroon | 18.7 | 84 | Sierra Leone | 32.2 |
| 8 | Albania | 6.3 | 53 | North Korea | 18.8 | 85 | Niger | 32.4 |
| 9 | Turkmenistan | 6.4 | 54 | Congo, Rep. | 19.1 | 86 | Burundi | 38.3 |
| 10 | Malaysia | 6.5 | | Andhra Pradesh | 19.5 | 87 | Eritrea | 39.0 |
| 11 | El Salvador | 6.5 | | Assam | 19.8 | 88 | Congo, Dem. Rep. | 42.7 |
| 12 | Morocco | 6.5 | 55 | Kenya | 19.9 | | | |
| 13 | Colombia | 6.7 | | Haryana | 20.0 | | | |
| 14 | South Africa | 6.9 | 56 | Sudan | 20.5 | | | |
| 15 | China | 7.1 | 57 | Nepal | 20.6 | | | |

Note: Only countries with GHI >5 are ranked.

Source: Table 2 and von Grebmer et al. 2008.

## THE ISHI IN RELATION TO OTHER SOCIAL AND ECONOMIC INDICATORS

How does the ISHI compare with other indicators of economic and social progress? As Figure 4 shows, the association between the hunger index and the percentage of the population below the poverty line is strong. This result is expected, given that poverty is often the root cause of insufficient food intake, child malnutrition, and child mortality; it also reflects the fact that the poverty line is developed using calorie cutoffs. A few states, however, deviate from the predicted line. Gujarat and Madhya Pradesh are clear "negative outliers," with a much higher hunger index than would be expected based on their poverty level. Punjab, Orissa, and Kerala, on the other hand, stand out as "positive deviants"—that is, they have significantly lower hunger index scores than would be expected of states with their level of poverty.

Figure 5 presents a scatter-plot of the 17 states by the hunger index and net state domestic product (NSDP) per capita, with the latter serving as a proxy for each state's per capita income. In this case, the two variables show a strong inverse association, with poorer states having a significantly higher hunger index than more prosperous states. The association is far from perfect, however, with a number of states appearing as outliers. For instance, Madhya Pradesh again stands out as having a much higher level of hunger than would be expected based on its per capita income; Jharkhand and Chhattisgarh are also "negative outliers," as is Maharashtra, which has a hunger index almost as high as that of Orissa but an NSDP twice as large. Several states are also doing better than expected given their economic level, with Punjab being a noticeable positive outlier, as well as, to a lesser extent, Kerala, Assam, and Rajasthan.

Finally, Figure 6 presents the association between the ISHI and the rate of economic growth for each state. The figure shows little evidence of a consistent relationship between the two variables. A state that experienced negative real growth (in net state domestic product per capita) between 1999–2000 and 2004–05 (for example, Madhya Pradesh) has a high hunger index, but so did states like Bihar, Jharkhand, and Chhattisgarh that experienced much higher rates of economic growth over this period. Again, Punjab stands out as a

**FIGURE 4—INDIA STATE HUNGER INDEX IN RELATION TO POVERTY**

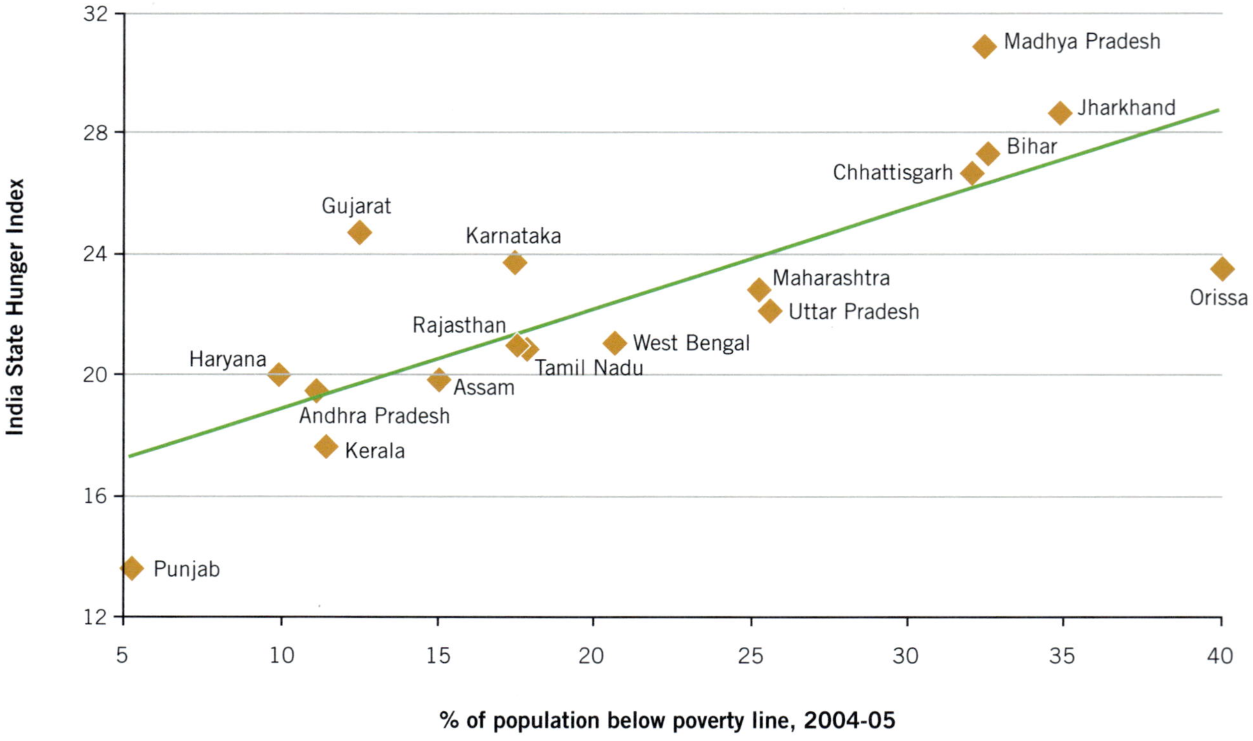


Source: Table 2 and www.indiastat.com.

FIGURE 5—INDIA STATE HUNGER INDEX IN RELATION TO PER CAPITA INCOME

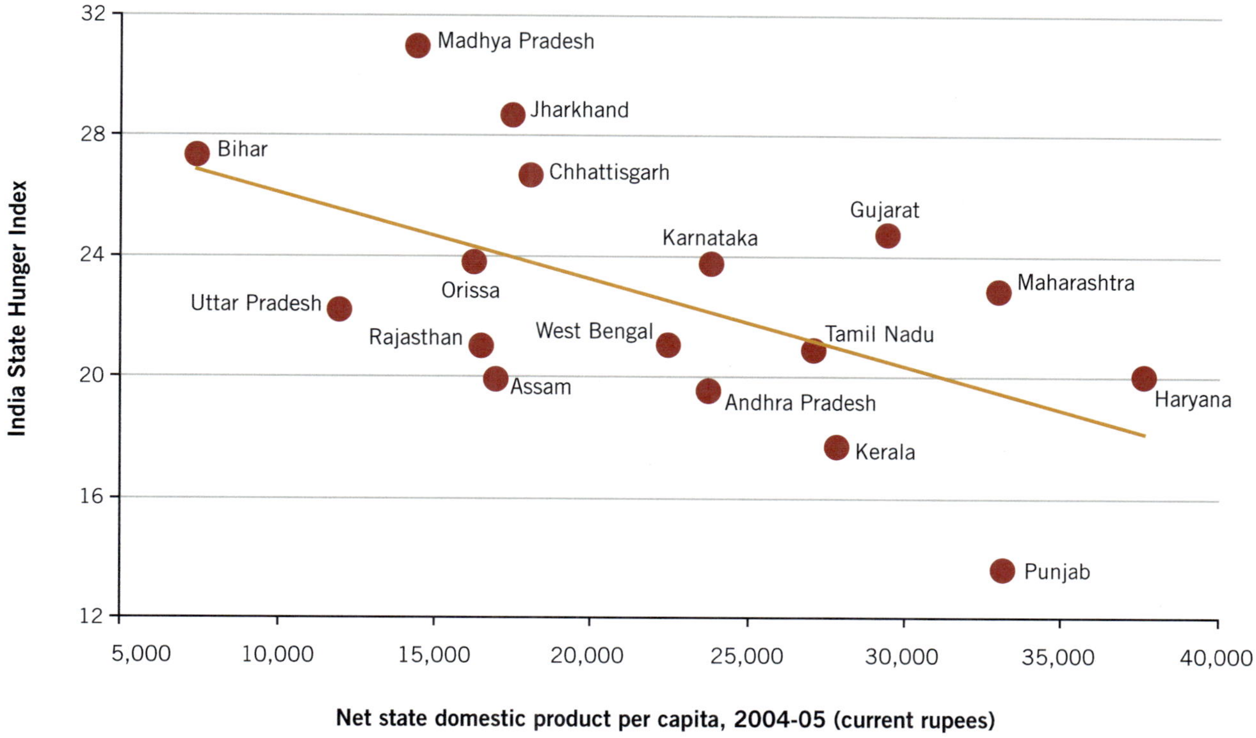


Source: Table 2 and www.indiastat.com.

FIGURE 6—THE INDIA STATE HUNGER INDEX IN RELATION TO ECONOMIC GROWTH

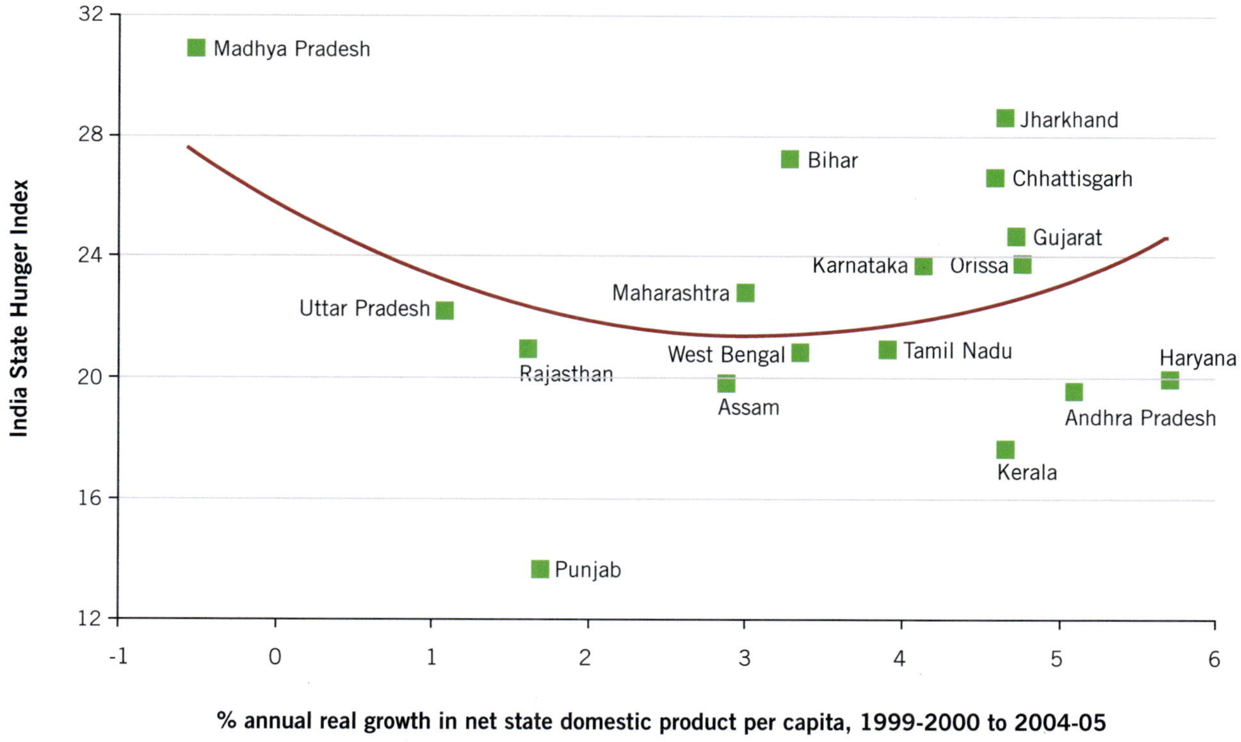


Source: Table 2 and www.indiastat.com.

remarkable positive outlier, with its much lower hunger index than states such as Kerala, Andhra Pradesh, and Haryana whose rate of economic growth was two to three times larger during the same period (<2 percent a year for Punjab, compared with 4 to 6 percent for the other three states). Thus, economic growth in this period appears to be weakly associated with a state's hunger index scores.

## TRENDS OVER TIME

Because the India State Hunger Index has not been estimated over two points in time, it is difficult to examine changes in the values of the state indexes over time. The similarity in construction of the India State Hunger Index to a state-level Nutrition Index[5] calculated using data for 1994 (Wiesmann 2004) gives us the ability to examine, at a minimum, the changes in rankings of the different states over the past 14 years.

Table 5 presents the rankings of states within India using the India State Hunger Index in 2008 and the Nutrition Index in 1994. Some distinct changes in ranking occurred over time. Particularly striking is the poor performance of the states of Orissa and Madhya Pradesh. In 1994, Madhya Pradesh ranked 11th out of the 15 states, whereas it ranked last (17th) in 2008. Given the large contribution of child underweight to the ISHI scores, the decline in ranking could be due to the lack of improvement in child undernutrition rates in Madhya Pradesh over the past seven years.[6] Similarly, Orissa, which ranked 5th on the Nutrition Index in 1994, now ranks 12th on the ISHI 2008. Haryana, despite its impressive economic performance, also seems to

[5] The Nutrition Index estimated by Wiesmann (2004) uses the same variables as the India State Hunger Index: child nutrition, child mortality, and calorie undernourishment. The Nutrition Index differs in that it uses FAO dietary energy supply data for estimating calorie undernourishment, and it uses prevalence of underweight among children under four years of age, rather than children under five years of age. Although index scores cannot be compared between the Nutrition Index and the ISHI, the comparisons are still valid from the point of view of comparing rankings across time.

[6] Authors' calculations based on the NFHS data show that the underweight rate among children under three years of age increased from 55.1 percent in 1998–99 to 57.7 percent in 2005–06.

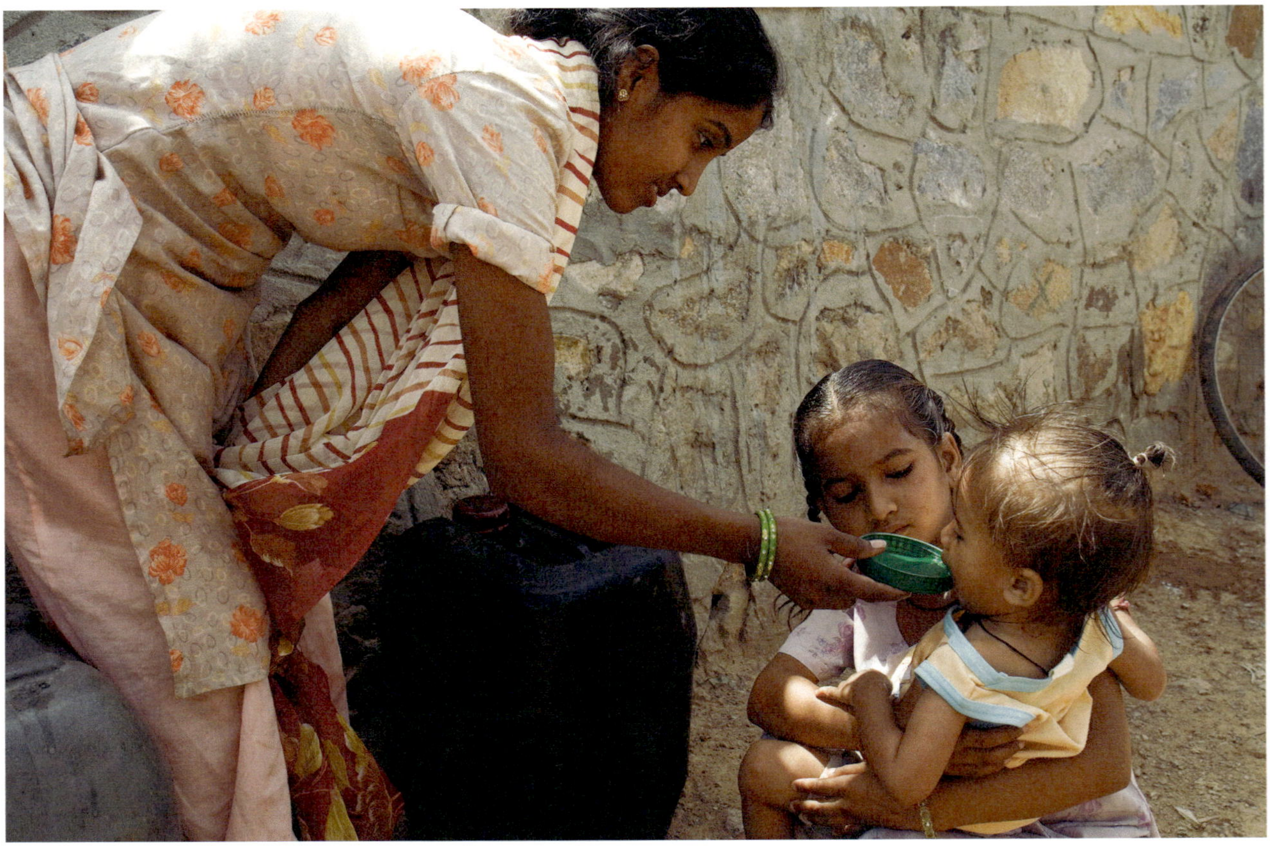

**TABLE 5—CHANGES IN STATE RANKINGS FROM THE NUTRITION INDEX 1994 TO THE INDIA STATE HUNGER INDEX 2008**

| Nutrition Index rank (1994) | State | India State Hunger Index rank (2008) | State |
|---|---|---|---|
| 1 | Haryana | 1 | Punjab |
| 2 | Kerala | 2 | Kerala |
| 3 | Rajasthan | 3 | Andhra Pradesh |
| 4 | Punjab | 4 | Assam |
| 5 | Orissa | 5 | Haryana |
| 6 | Andhra Pradesh | 6 | Tamil Nadu |
| 7 | West Bengal | 7 | Rajasthan |
| 8 | Uttar Pradesh | 8 | West Bengal |
| 9 | Karnataka | 9 | Uttar Pradesh |
| 10 | Gujarat | 10 | Maharashtra |
| 11 | Madhya Pradesh | 11 | Karnataka |
| 12 | Tamil Nadu | 12 | Orissa |
| 13 | Maharashtra | 13 | Gujarat |
| 14 | Assam | 14 | Chhattisgarh |
| 15 | Bihar | 15 | Bihar |
| | | 16 | Jharkhand |
| | | 17 | Madhya Pradesh |

Note: Nutrition Index 1994 results are from Wiesmann (2004).

have performed poorly in terms of reduction of hunger; although it was ahead of all other states in 1994, it now ranks 5th on the ISHI 2008.

A few states have outperformed others in enhancing food and nutrition security, and this improvement is apparent in upward changes in the ranks between the Nutrition Index and the ISHI 2008. Noteworthy among these is the increase in the ranking of Assam. The state was one of the poorest performers and ranked 14th on the Nutrition Index in 1994, but became one of the best performers in 2008, in spite of having had the lowest growth rate of per capita income over the past 14 years among all states. Tamil Nadu also performed well and improved its ranking from 12th in 1994 to 6th in 2008.

Overall, the changes in ranking are somewhat sobering. The trends in the few states that have improved despite the economic odds underscore the importance of investments in social protection, health, and nutrition services to ensure progress on poverty and hunger alleviation. Continued monitoring of trends using indicators like the India State Hunger Index is essential to monitor progress and attract attention to the issue of hunger and undernutrition.

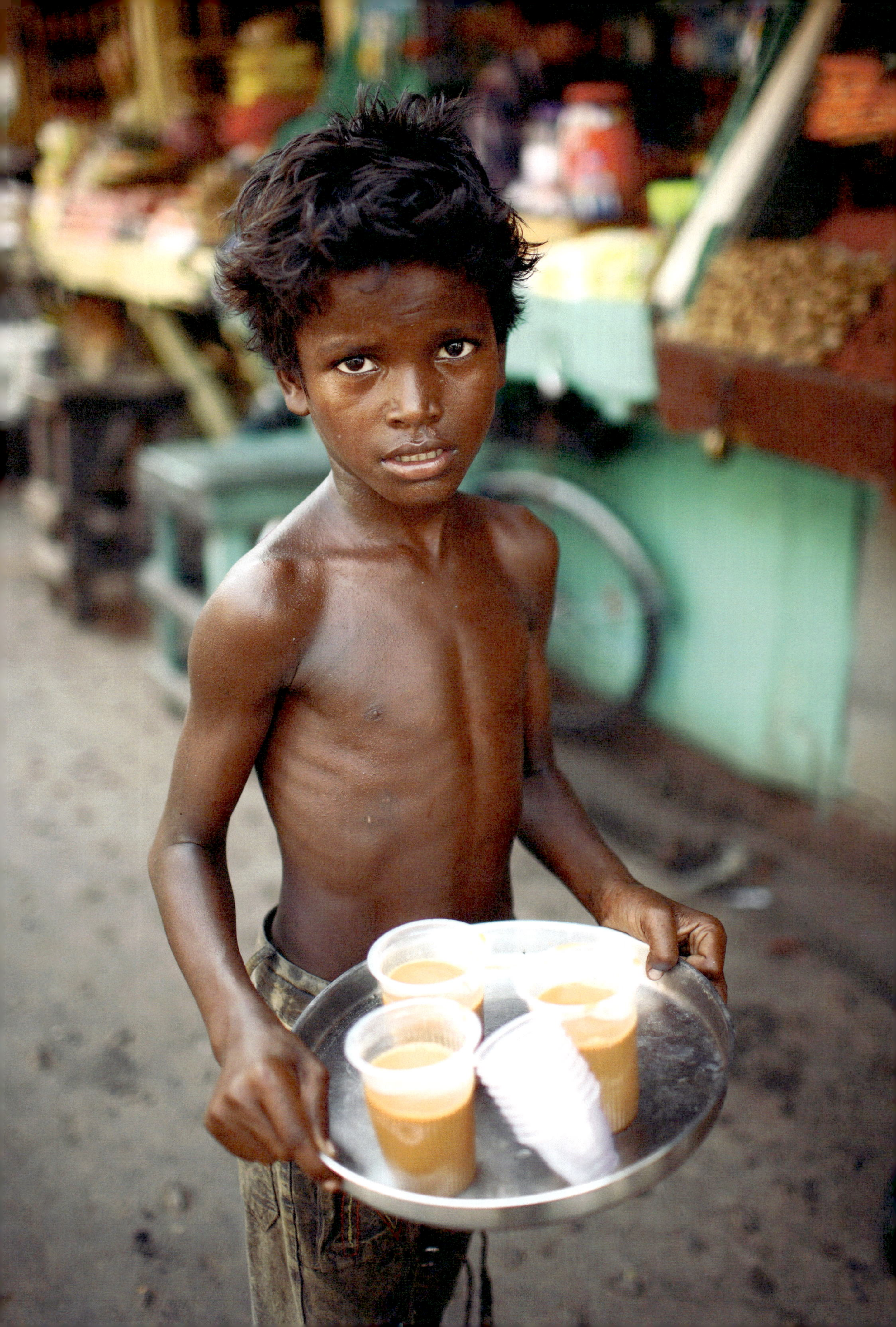

# Summary and Policy Implications

The results of the India State Hunger Index 2008 highlight the continued overall severity of the hunger situation in India, while revealing the variation in hunger across states within India. It is indeed alarming that not a single state in India is either low or moderate in terms of its index score; most states have a "serious" hunger problem, and one state, Madhya Pradesh, has an "extremely alarming" hunger problem.

Although variation exists in index scores of the states, and hence in the ranking of Indian states in relation to other countries, few states perform well in relation to the GHI 2008. Even the best-performing Indian state, Punjab, lies below 33 other developing countries ranked by GHI. Even more alarming is the fact that the worst-performing states in India—Bihar, Jharkhand, and Madhya Pradesh—have index scores similar to countries that are precariously positioned on the GHI 2008 rankings. For instance, Bihar and Jharkhand rank lower than Zimbabwe and Haiti, whereas Madhya Pradesh falls between Ethiopia and Chad.

Our analysis of the associations between the ISHI 2008 and state economic indicators shows that the relationship between poverty and hunger is largely as expected—greater ISHI 2008 scores are seen in poorer states, with a few exceptions. Outliers like Kerala, Orissa, and Punjab perform better on the ISHI 2008 than might be expected given their poverty levels, whereas Gujarat, Karnataka, and Madhya Pradesh perform worse. A closer examination of these states' past and current investments in social protection, health, and nutrition programs can help inform the debate about policy instruments to protect populations against hunger even in the face of poverty.

The lack of a clear relationship between state-level economic growth and hunger, taken along with the relationship between the ISHI 2008 and poverty and incomes, has a number of implications. First, economic growth is not necessarily associated with poverty reduction. Additionally, even if equitable economic growth improves food availability and access, it might not lead immediately to improvements in child nutrition and mortality, for which more direct investments are required to enable rapid reductions. Thus, in addition to wide-scale poverty alleviation, direct investments in improving food availability and access for poor households, as well as direct targeted nutrition and health interventions to improve nutrition and mortality outcomes for young children, will be needed to raise the ISHI scores and rankings of Indian states.

Child underweight contributes more than either of the other two underlying variables to the GHI score for India and to the ISHI scores for almost all states in India. Tackling child undernutrition, therefore, is crucially important for all states in India. Achieving rapid reductions in child underweight, however, will require scaling up delivery of evidence-based nutrition and health interventions to all women of reproductive age, pregnant and lactating women, and children under the age of two years.

Some economically strong states had rankings on the Nutrition Index that deteriorated when compared with the ISHI 2008, suggesting that it might be important for these states to invest in direct nutrition and poverty alleviation interventions even during sustained economic growth. The design and implementation of policies and programs to improve all three underlying dimensions of the ISHI will need to be strengthened and supported to ensure that

hunger is reduced rapidly over time. Although strides are being made on the public health front to ensure sustained reductions in child mortality, improvements in child nutrition are not satisfactory in India. Nutrition programs in India are not effectively delivering evidence-based interventions at scale to vulnerable age groups that need to be reached to ensure rapid reductions in undernutrition.

In conclusion, for Indian states to progress along the ISHI, and to ensure that ISHI scores for Indian states are more closely aligned with GHI scores of countries with comparable economic growth, investments will be needed to strengthen agriculture, improve overall food availability and access to all population segments, and to improve child nutrition and mortality outcomes.

# References

Dandekar, V. M., and N. Rath. 1971. *Poverty in India.* Indian School of Political Economy, Pune.

Deaton, A., and J. Drèze. 2008. Nutrition in India: Facts and interpretations. Mimeo. Available at Social Science Research Network, http://ssrn.com/abstract=1135253.

Dev, S. M. 2005. Calorie norms and poverty. *Economic and Political Weekly,* February 19, 789–792.

FAO (Food and Agriculture Organization of the United Nations). 2008. The state of food insecurity in the world. Rome.

GOI (Government of India). 1979. Report of the Task Force on Projections of Minimum Needs and Effective Consumption Demand. Planning Commission, New Delhi. Mimeo.

Gopalan, C., B. V. Ramasastry, and S. C. Balasubramanian, revised and updated by B. S. Narasinga, Y. G. Deosthale, and K. C. Pant. 1991. *Nutritive values of Indian foods.* New Delhi: Indian Council of Medical Research.

IIPS (International Institute of Population Sciences). 2007. *National Family Health Survey 2005–06 (NFHS-3), National Report.* Mumbai: IIPS.

NSSO (National Sample Survey Organisation). 2007. Nutritional intake in India 2004–05. Report 513 (61/1.0/6). Ministry of Statistics and Programme Implementation, Government of India, New Delhi. Mimeo.

Smith, L. C., and D. Wiesmann. 2007. *Is food insecurity more severe in South Asia or in Sub-Saharan Africa? A comparative analysis using household expenditure survey data.* IFPRI Discussion Paper 712. Washington, DC: International Food Policy Research Institute.

United Nations. 2006. *The Millennium Development Goals report 2006.* New York.

von Grebmer, K., H. Fritschel, B. Nestorova, T. Olofinbiyi, R. Pandya-Lorch, and Y. Yohannes. 2008. *Global Hunger Index report 2008.* Bonn, Washington, DC, and Dublin: Welthungerhilfe, International Food Policy Research Institute, and Concern.

WFP and MSSRF (World Food Programme and M. S. Swaminathan Research Foundation). 2002. *Food insecurity atlas of rural India.* Chennai, India: MSSRF.

Wiesmann, D. 2004. *An international nutrition index: Concept and analyses of food insecurity and undernutrition at country levels.* Development Economics and Policy Series No. 39. Frankfurt: Peter Lang.

———. 2006. *A global hunger index: Measurement concept, ranking of countries, and trends.* Food Consumption and Nutrition Division Discussion Paper 212. Washington, DC: International Food Policy Research Institute.

Wiesmann, D., L. Weingartner, I. Schoeninger, and V. Schwarte. 2006. *The challenge of hunger: Global Hunger Index: Facts, determinants, and trends. Case studies in the post-conflict countries of Afghanistan and Sierra Leone.* Bonn and Washington, DC: Welthungerhilfe and International Food Policy Research Institute.

Wiesmann, D., A. K. Sost, I. Schoeninger, H. Dalzell, L. Kiess, T. Arnold, and S. Collins. 2007. *The challenge of hunger 2007: Global Hunger Index: Facts, determinants, and trends. Measures being taken to reduce acute undernourishment and chronic hunger.* Bonn, Washington, DC, and Dublin: Welthungerhilfe, International Food Policy Research Institute, and Concern.

World Bank. 2007. *World development indicators 2007.* Washington, DC.

# Appendix: Data Appendix

Data on under-five mortality rates and child underweight rates for each state were obtained directly from the published report of the third National Family Health Survey (IIPS 2007). As noted in the text, the proportion of the population consuming inadequate calories was calculated directly using detailed household-level food consumption data from the 61st round of the National Sample Survey, conducted in 2004–05. The NSS obtained household consumption data on more than 225 individual foods; these were converted to calories using food-to-calorie conversion factors reported by NSSO (2007). The conversion factors used by the NSSO are largely based on the article "Nutritive Values of Indian Foods" (Gopalan et al. 1991).

The NSS data include information on the number of meals consumed by household members at home and away from the household (either free or on payment). Also included is information on meals consumed in the household by guests and employees. According to the NSSO recommendation, household nutrient intake derived from the detailed food consumption data should be adjusted up or down by the scaling factor $\alpha$, which is given by

$\alpha = (M_h + M_f)/ (M_h + M_g + M_e)$, where

$M_h$ = number of meals consumed by household members at home,

$M_f$ = number of free meals received by members as guests or employees,

$M_g$ = number of meals consumed at home by guests, and

$M_e$ = number of meals consumed at home by household employees.

Thus, the "true" measure of household nutrient intake would be

$$C = \alpha C_d ,$$

where $C_d$ is the derived level of consumption from the detailed food consumption data.

We used this methodology to adjust the derived level of household calorie intake. To calculate calorie intake per person per day, we divided the "true" household monthly calorie intake by 30 days and by the number of members residing in a household.

## RATIONALE FOR THE USE OF THE 1,632-KCAL CUTOFF TO CALCULATE THE PROPORTION OF THE POPULATION UNDERNOURISHED

To calculate the calorie-deficient population, we used a calorie cutoff of 1,632 kcals per person per day and then calculated the weighted proportion of households consuming fewer than this level of calories per person per day. As explained in the text, 1,632 kcals is the level of daily calorie intake per capita that results in exactly 20 percent of the Indian population being calorie-deficient—the level of calorie undernutrition estimated by the FAO using data from food balance sheets. Since an important objective of this study is to be able to compare individual Indian states with other developing countries, it was important for us to use data and a methodology that would result in roughly consistent calorie-deficiency figures at the all-India level using either the ISHI or the GHI. The use of the 1,632-kcal cutoff ensures this consistency.

Figure A.1 shows the cumulative distribution of mean daily calorie intake per capita in the country as a whole using the NSS unit-level data. The straight vertical line represents the cutoff of 1,632 kcals per person per day. The figure also shows that 20 percent of the population falls below this cutoff. Not surprisingly, the cumulative distribution is steep around this point, so the proportion of the population that is calorie-deficient is very sensitive to small changes in the cutoff. For instance, shifting the calorie cutoff to just 1,820 or 2,000 kcals raises the calorie-deficient population to 34 or 48 percent, respectively.

**FIGURE A.1—CUMULATIVE DISTRIBUTION OF AVERAGE DAILY CALORIE INTAKE PER CAPITA, INDIA, 2004–05**

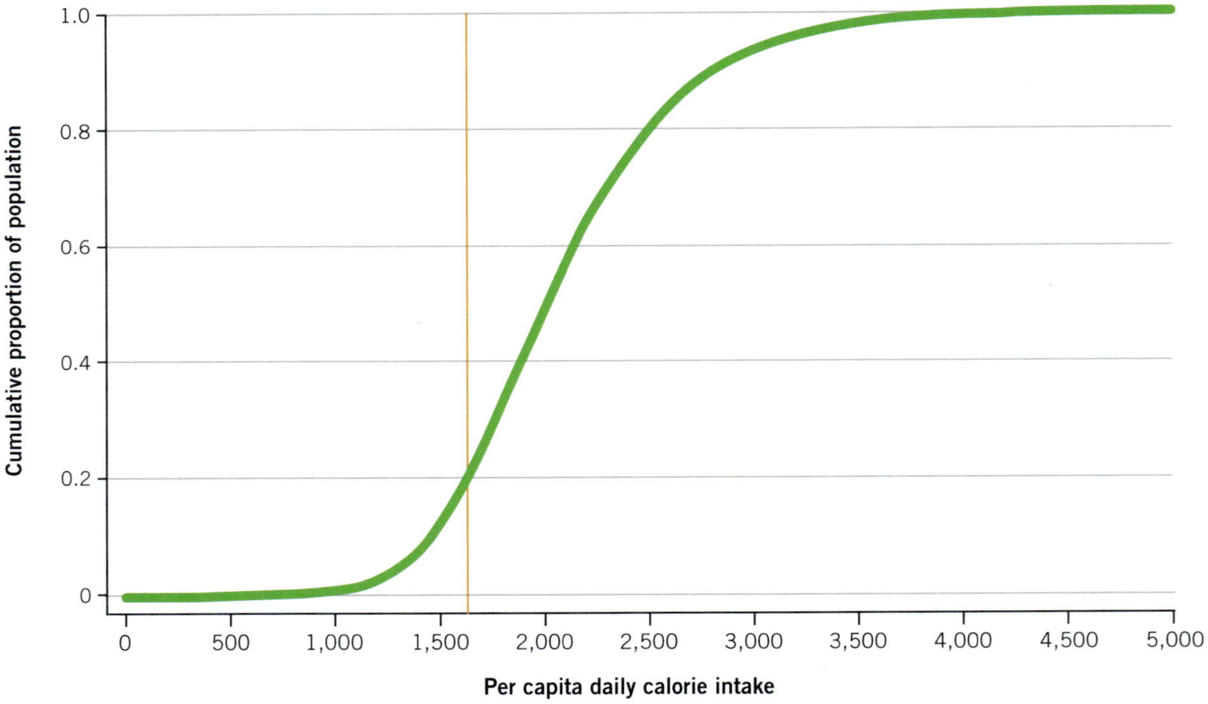


How do our calorie-deficiency figures compare with those reported by the NSS itself or by other researchers? Unfortunately, the NSS report uses a cutoff level of 2,700 kcals per consumer unit (or adult equivalent) per day. Comparison is thus difficult, for calorie intake per person and calorie intake per consumer unit are very different variables, and their use can yield very different calorie-inadequacy numbers. The NSS reports that approximately 61.3 percent of the rural population and 63.1 percent of the urban population consume fewer than 2,700 kcals per consumer unit per day.

Deaton and Drèze (2008) recently wrote a paper on undernutrition in India using data from the 61st round of the NSS to analyze hunger and malnutrition in India. Although they do not report the cumulative distribution of daily calorie intake per capita in their paper, they do report that average calorie intake per person per day is 1,624 kcals for the bottom quartile of the Indian population (ranked by per capita household consumption expenditure). This number is broadly consistent with our figure of 20 percent of the population consuming 1,632 or fewer kcals per person per day.

Given that all three sources—the NSSO (2007), Deaton and Drèze (2008), and ourselves—use the same unit-level data from the NSS 61st round survey and apply the same methodology to calculate the caloric value of foods, it is obvious that the three sources will come up with the same proportion of calorie-deficient population. The more interesting question is why the FAO food balance sheets show that a much lower proportion of the population is undernourished. For instance, the FAO reports that 20 percent of the Indian population has access to 1,820 or fewer kcals per person per day. Using the same calorie cutoff, the NSS data indicate that 33.8 percent of the population is undernourished.

Obviously, not all of the food supply is consumed by humans. A considerable amount is used for animal feed, and some is lost due to spoilage, shrinkage, and transportation and storage losses, among other things. In calculating food availability per capita in its food balance sheets, the FAO makes certain assumptions about the proportion of the food supply that is unlikely to result in direct human consumption. A comparison of the FAO and NSS data for India suggests, however, that the FAO may have underestimated the amount of food loss and food going to animal consumption, since the NSS data indicate calorie consumption levels that are approximately 10 percent lower than those indicated by the FAO food balance sheets. Of course, it is possible that the NSS data have underestimated calorie consumption, but, in general, data obtained from a direct household survey are likely to be more reliable than secondary information obtained from public food production statistics. In our case, we had no choice but to use the NSS household survey data, because the FAO obtains and reports food balance sheets only for India as a whole, not for individual states within the country. Because our main objective in this report is to come up with a state-level hunger index, we cannot use the FAO food balance data to derive state-specific levels of calorie inadequacy.

Finally, Figure A.2 shows the cumulative distribution of per capita daily calorie consumption for two states—Punjab and Tamil Nadu—to demonstrate the large variations in calorie consumption that are found within India. These two states represent the minimum and maximum levels of calorie deficiency within our sample of 17 major states. The entire calorie distribution curve for Punjab is significantly to the right of that for Tamil Nadu, with the result that a much larger proportion of individuals in Tamil Nadu (29.1 percent) is undernourished than in Punjab (11.1 percent), based on this indicator.

**FIGURE A.2—CUMULATIVE DISTRIBUTION OF AVERAGE DAILY CALORIE INTAKE PER CAPITA, PUNJAB AND TAMIL NADU, 2004–05**

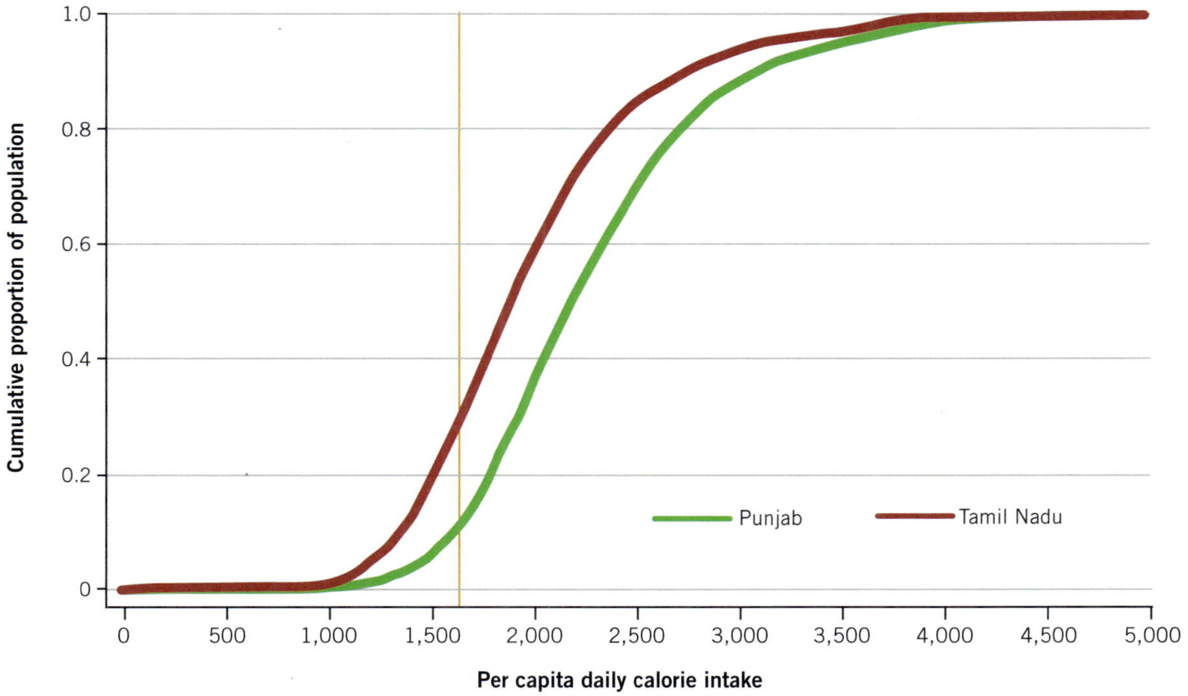

## PHOTO CREDITS

Cover photo: © iStockphoto.com/Kamruzzaman Ratan
Page 6: © Panos Pictures/Abbie Trayler-Smith
Page 8: © Panos Pictures/Atul Loke
Page 9: © Panos Pictures/Zack Canepari
Page 10: © Panos Pictures/G.M.B. Akash
Page 12-13: © Panos Pictures/Atul Loke
Page 14: © Panos Pictures/Tom Pilston
Page 22: © Panos Pictures/Stuart Freedman
Page 24: © Panos Pictures/G.M.B. Akash
Page 26: © Panos Pictures/Jacob Silberberg
Page 28: © Panos Pictures/Jacob Silberberg
Page 32: © Panos Pictures/G.M.B. Akash

Made in the USA
Monee, IL
05 November 2026

67379790R00021